Mohamed Al-saadi

The Smart Power Strip

Mohamed Al-saadi

The Smart Power Strip

LAP LAMBERT Academic Publishing

Imprint
Any brand names and product names mentioned in this book are subject to trademark, brand or patent protection and are trademarks or registered trademarks of their respective holders. The use of brand names, product names, common names, trade names, product descriptions etc. even without a particular marking in this work is in no way to be construed to mean that such names may be regarded as unrestricted in respect of trademark and brand protection legislation and could thus be used by anyone.

Cover image: www.ingimage.com

Publisher:
LAP LAMBERT Academic Publishing
is a trademark of
Dodo Books Indian Ocean Ltd. and OmniScriptum S.R.L publishing group

120 High Road, East Finchley, London, N2 9ED, United Kingdom
Str. Armeneasca 28/1, office 1, Chisinau MD-2012, Republic of Moldova, Europe
Printed at: see last page
ISBN: 978-3-659-91912-1

The Smart Power Strip

Done By: Eng. Mohammed Omar ALSaadi

Electrical Engineer

TABLE OF CONTENTS

Chapter One: Introduction to Electricity and electric energy

1. INTRODUCTION 6

2. GENERATION OF ELECTRICITY 6

3. SOURCE OF ENERGY 8
   3.1 Renewable Energy Sources 8
   3.2 Nonrenewable Energy Sources 9

4. CONSERVATION OF ELECTRICITY 10

Chapter Two: Introduction to energy measurement and labeling

1. INTRODUCTION 13

2. ENERGY STANDARDS 13

3. MEASUREMENT SYSTEM 15

4. ENERGY LABELS 15

Chapter Three: Energy labeling

1. INTRODUCTION 18

2. DEFINITION 20
   2.1 endorsement and comparison 20

2.2 comparison labels basic types 21

3. IMPORTANCE OF ENERGY LABEL 22
3.1. Importance for Retailers 22
3.2. Importance for a Client 22
3.3. Importance for Manufacturers 22

4. ENERGY LABEL SCHEMATICS 22
4.1. Information Content of the Energy Label 24
4.2. Energy labeling examples 25

5. SUMMARY 26
5.1. Arguments for Customers 26
5.2. Energy Labeling = Legal Duty 26
5.3. Saving energy at home 26

Chapter Four: Calculation and design part

1. INTRODUCTION 28
1.1 Objectives 28
1.2 Block Diagram 29

2. DESIGN PROCEDURE / DESIGN DETAILS 30
2.1 Schematics 30
2.2 Main Protection 30
2.3AC/DC Circuit 32
2.4 Microcontroller 33
2.5 Current Sense 36

3. DESIGN VERIFICATION 38
3.1 Testing Main Protection 38
3.2 Testing AC/DC Circuit 39
3.3 Testing Current Sense 41

4. CONCLUSIONS 42
4.1 Accomplishments 42
4.2 Uncertainties 42

5. REFERENCES 43

CHAPTER ONE: INTRODUCTION TO ELECTRICITY AND ELECTRIC ENERGY

1. Introduction:

Energy is defined as "the ability to do work." In this sense, examples of work include moving something, lifting something, warming something, or lighting something. Electricity is one of the most powerful forces in our lives. As a matter of fact, it can even kill you. The most vital part of electricity is called electric energy. This is what we commonly think of when we hear the word electricity. “Electricity” reminds us of anything that we plug into an electrical outlet in order to make it work, such as lights, refrigerators, video games, microwaves, and computers. Scientists discovered ways to produce electric energy in large amounts in order to make heat, light and motion. These discoveries have improved our lives greatly and for many of us it would be difficult to picture what life would be like without electricity.

2. Generation of electricity:

Everything in the world, including air, humans, water, etc. is made up of atoms, or tiny invisible particles. Protons and neutrons join together to form the nucleus or center of the atom. Electrons, which are even smaller, spin around the nucleus of the atom. When electrons move from one atom to another, they form electricity. So, how do these protons and neutrons help make our computer work at home?

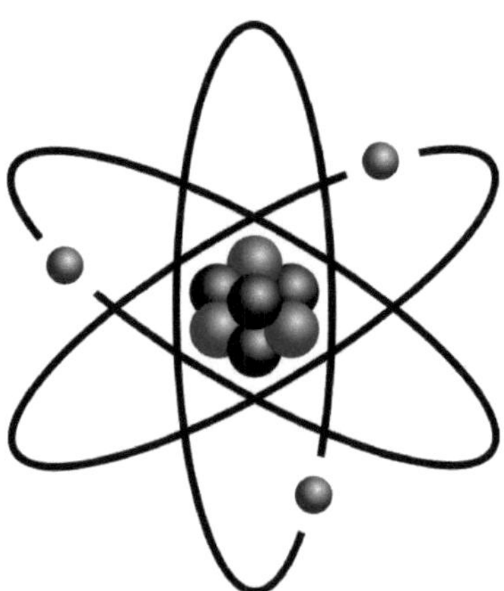

In today's world, huge generators at power plants make electricity by using coal, natural gas, uranium, water, or wind. Electric current travels through transformers which push the power a long distance through transmission lines that run across countries. The electric current often goes through a substation and then travels on smaller power lines or along underground cables through wires in the neighborhood. A service drop is a pole that connects the power line to your house. The current runs through a meter that adds up how much electricity your family uses. Then the electricity goes into a service panel in your basement or garage. Finally, the electrical current moves through the wires in the wall of your house to all of the outlets and switches in your house. Now you know why your computer will work when you turn it on.

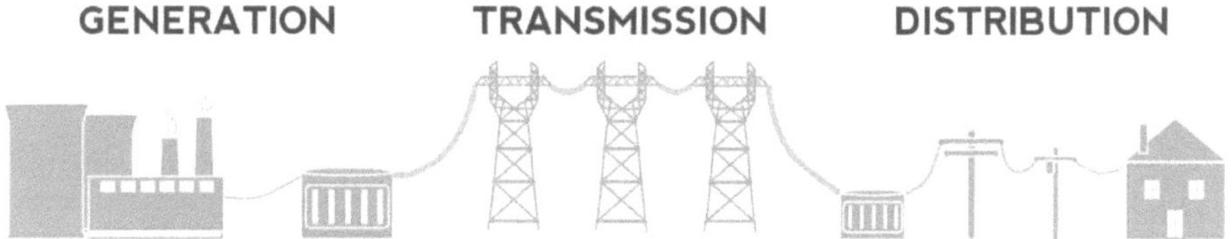

3. Sources of energy:

-Renewable energy sources

-Nonrenewable energy sources

3.1 Renewable energy sources:

These sources are constantly renewed or restored and include wind (wind power), water (hydropower), sun (solar), vegetation (biomass), and internal heat of the earth (geothermal).

3.2 Nonrenewable energy sources:

These are natural resources that cannot be replenished (fossil fuels such as oil, gas, and coal). In addition to renewable and nonrenewable energy sources, electric power is generated by nuclear power plants. However, operating such plants poses significant nuclear waste disposal problems; consequently, there are no current plans to build more. **Most electricity is generated by burning nonrenewable fossil fuels and there is a limited amount of these energy sources.**

4. Conservation of electricity:

Every year, many industrialized countries around the world use more energy than they did the year before. Some experts say that the total amount of energy use doubles every twenty years. One of the biggest problems is that a lot of this energy is wasted.

There are many reasons why we need to conserve, or use less, electricity. First of all, if we conserve electricity, we save money because electricity costs money. If you conserve electricity every single day, you can save money over time. You can spend this money on other things that you need and want. Second, conserving electricity means that we are using less of the earth's non-renewable resources. Non-renewable resources are gone forever once consumed. Examples are oil, natural gas, uranium, and coal which are used to create electricity. Therefore, wasting electricity is not good for the environment.

Another reason that we need to conserve energy is to make up for some of the energy that is wasted. . Companies use electricity in the production of items. At the same time, consumers do not reuse, neither recycle the produced items, instead throwing it away which increases the accumulation of waste and garbage. Within this process from manufacturing to consumption, a lot of energy and electricity is wasted.

Additionally, sulfur dioxide is also emitted into the air when coal is burned. The sulfur dioxide reacts with water and oxygen in the clouds to form precipitation known as "acid rain." Acid rain can kill and harm living creatures such as fish and trees and damage limestone buildings and statues.

Finally, we need to conserve electricity today in order to help future generations. Energy will be even a bigger problem in the future than it is today. So if younger generation starts learning to use less electricity and use it wisely today, they will be helping themselves in the future.

CHAPTER TWO: INTRODUCTION TO ENERGY MEASUREMENT AND LABELING FOR APPLIANCES AND EQUIPMENT

1. Introduction:

National energy efficiency measurement and monitoring became an important component of energy strategy in many countries, especially the energy-deficient ones, in the aftermath of the 1973 world oil crisis. With substantial increases in world oil prices, many countries recognized the need to understand how effective energy was being consumed in their economies and to increase energy efficiency. To serve these purposes, appropriate energy efficiency indicators were developed and applied so that any efficiency changes that took place could be quantitatively expressed. These indicators were also used in cross-country comparisons to explain differences in energy performance between countries and for international benchmarking.

Since the late 1980s, resulting from the growing concern about global warming caused by the burning of fossil fuels, energy efficiency improvement has become yet another natural step for countries to take to reduce greenhouse gas (GHG) emissions. In particular, at the 1997 third Conference of the Parties held in Kyoto, participating countries agreed to a timetable of GHG emissions reductions for the years 2008–2012 relative to the 1990 levels. A key element of the strategy of most countries to meet their reduction objective is to take steps to increase energy efficiency in all sectors of the economy. It has become necessary for them to assess fulfillment of energy efficiency improvement targets in a rigorous manner. As a result, there has been new emphasis on the development and application of economy-wide energy efficiency indicators both for evaluation and monitoring purposes.

2. Energy standards:

Standards are a set of procedures and regulations that prescribe the energy performance of manufactured products, sometimes prohibiting less energy efficient than the minimum standard.

Energy efficiency standards can be either mandatory or voluntary. They can be in the form of eight minimum allowable energy efficiency or a maximum allowable energy use. A mandatory energy efficiency standard is generally the most effective means of rapidly improving the energy efficiency of appliances. While, voluntary energy efficiency standards negotiated between government and manufacturers is an alternative option to energy efficiency standard. They have merit of being less controversial and hence some easier to enact.

The statistical approach requires fewer data and less analysis than the engineering economic approach. The data required are those which give a current characterization of marketplace for the products of interest. This approach looks at the models available at a particular time and performs an egression analysis to determine the dependence of energy use on adjusted capacity. Using such approach, policy makers can decide on the percentage of models they are willing to eliminate or the desired overall energy savings from the standards. After calculation of the regression line, the least energy efficient model is found and replaced with a model of higher efficiency. The energy savings of replaced model are calculated and energy savings are aggregated until the total reaches the desired goal. The minimum efficiency line is defamed as the line of maximum efficiency index. The efficiency index of a model is the percentage that the energy is above or below the reference line depends on efficiency indication. This approach has been utilized in the European Union (EU) and in Australia.

The second approach is engineering economic analyses. This approach has been widely used by Lawrence Berkeley Laboratory (LBL) for the U.S. department of energy. An engineering analysis is carried out on each product type to determine manufacturing costs for improving the efficiency of auricular model. Following seven steps performs the basis for an engineering analysis:

(i) Selection of appliance classes (ii) Selection of particular units (iii) Selection of design options for each class (iv) Calculation of efficiency improvement from each design option (v) combination of design options and calculation of efficiency improvements (vi) developing cost estimates for each design option (vii) generation of cost-efficiency curves.

Nevertheless, whatever methodology in establishing efficiency standards, the following circumstances should be taken into consideration:

(i) The level of the standard must have a positive effect on the environment.

(ii) Before implementing the standard, consumer should be protected against the high rise in total costs over the life of the given appliance.

(iii) The standard should ensure energy efficiency in relation to performance and should not affect the quality of the appliance.

(iv) The standard should ensure market competitiveness.

Finally, it is necessary to ensure that the efficiency standards are dynamic, so that they do not end up static in their effort to make electricity consumption efficient. It is possible to maintain their dynamism by creating gradual standards that in certain ranges make increasing demands on electricity consumption.

3. Measurement system:

Many of the problems identified in Power scope can be avoided by building a computer with hardware support for taking the measurements. One of these problems is, for example, the attribution of background activities, such as network traffic or hard-disk data transfer, to the wrong process.

By splitting up the measurements into separate entities for these devices and associating those with modules instead of processes, a deferred attribution to processes is possible. For example, a system with a CPU, memory, network card, and hard disk, would measure the power consumed by each of these items independently. The system can then associate the CPU measurement with the running process, the network card with processes which have submitted or received packets, the disk with processes using file systems, and the memory system with processes causing memory bus accesses (which may be associated with devices, owing to direct memory access (DMA)).

4. Energy Labels:

The purpose of introducing labels is to convince consumers to buy and manufactures to produce more efficient appliances. Labels should enable the comparison of energy efficiency or cost for similar appliances that compete with types having similar dimensions and characteristics. In considering a purchaser can see the price and choose the product with the lowest long-term costs. The indication energy label concept can be divided into three types: i) Labels with energy costs: this type of label is usually used for refrigerator, freezer, water heater, dishwasher, and washing machines. Energy cost for one year of operation are listed for the average price of energy and for the price offered by the local distribution company. An indication comparing the types and models is also mandatory, the minimum and maximum in corresponding category. ii) Labels with energy efficiency: this type of label is used generally for air conditioner, they compare energy efficiency or the price of energy for a certain number of hours of operation. The label must include a graph comparing the product to other types as well as information on how to use the product most efficiently. iii) Labels with general information: this type of label is usually use for ovens and small boiler. They primarily include information about efficient methods of home heating and ways to use heating devices in order to achieve maximum energy efficiency.

For maximum effect of labeling, it is necessary to adhere to several fundamental conditions:

(i) The label must be uniform, if labeling or information flyers are introduced in various ways, it can cause chaos in comparing appliances and lead the consumer to ignore information about energy consumption.

(ii) Label must be general, and all appliances of a given type must be labeled. If they are not, there is a danger that consumers would not want to know the operational cost. Thus they would give priority to less effective appliances without labels over effective appliances with labels.

(iii) The labels and information flyer must be as exact as possible while giving as much necessary information as possible.

(iv) Measure label should include informational flyer about the product for those customers who are willing to devote more time to considering the relative benefits of various appliances.

(v) Energy labels measures should allow informational flyer to eliminate all obstacles hindering trade between member countries. The effort should be to support a unified market.

CHAPTER THREE: ENERGY LABELING

1. Introduction:

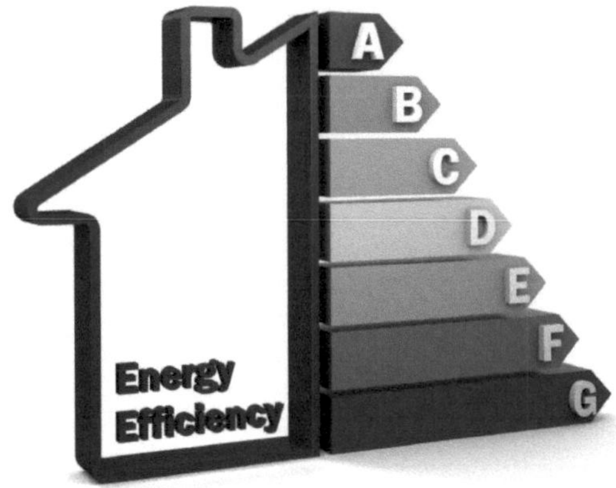

The world today faces a daunting list of environmental issues. The "greenhouse" effect is one, and has been described as the most potentially damaging problem that faces the world today, with about half the UK`s emission of the major greenhouse gas, carbon dioxide, attributed to buildings. Currently the energy used in housing gives rise to about 175 million tons of carbon dioxide emissions per year. This represents about 8 tons per house. Energy efficient dwellings give rise to significantly smaller emissions. The need for housing to consume less fuel is therefore of imperative importance. Improving the energy efficiency of housing will lead to lower fuel use and a subsequent reduction of emissions together with improved comfort levels through affordable heating.

Domestic housing uses energy inefficiently. Its construction is an area where much improvement could be made in terms of energy conservation. Legislation in the UK sets standards for the thermal conductivity of materials used within the construction of new dwellings, requirements for the insulation of dwellings being first introduced into the Building Regulations of 1965. The current Building Regulations provide a minimum standard 170 of limiting energy loss, (often interpreted as a maxima by developers and builders when trying to minimize construction costs). These regulations take modest account of air infiltration, cold bridging, and the management of energy use.

As levels of fabric insulation increase, other heat-loss factors become more important. Limiting heat loss by reducing ventilation increases the risk of condensation.

Providing mechanical ventilation with heat recovery diminishes this risk whilst recouping some of the running costs. Heat recovery systems are not however, considered to be cost-effective in dwellings with efficient heating systems, as the electricity to run them becomes more critical. It is important to provide adequate and controllable ventilation and reduce air leakage through adventitious routes. Cold bridging becomes more significant with higher fabric insulation standards - even the nails in timber framing can become potential heat-loss paths. Energy use is probably a matter for education of building users. Legislation has made no attempt to address this aspect of energy use.

Buildings can be designed to minimize the use of artificial heat energy, through passive solar heating, improved insulation and construction, massing of the construction, more efficient appliances and lighting. The question remains, do we currently do enough to conserve and utilize energy, or should we compel designers to incorporate energy efficient measures within new buildings. How far we can impose energy efficient measures on housing is an immediate and taxing problem facing the constructors of building legislation.

The inclusion of energy labelling into the Building Regulations is a further improvement that could also act as a possible marketing tool to attract house-buyers. Providing an overall energy rating for a house could require designers to consider further measures such as orientation, passive solar gains, glazing strategy, and energy efficient heating systems, in order to maximize the energy rating.

Limited work has been carried out on public attitudes to energy conservation. The perceived value of energy labelling has not been tested. Part of the purpose of this work is to assess the views of designers on the usefulness of energy labelling.

Benefits:

- Energy labels contain basic information about energy consumption (as well as water consumption) and shows the energy class
- It enables the calculation of the overall operation costs of the appliance
- The law prescribes that an energy label has to be shown on the appliance at the point of its sale.

2. Definition:

It is axiomatic that the market for household energy services would be enhanced where buyers are able to take into account not just the cost of the appliance but the otherwise invisible factor of energy consumption. Energy labels improve the market's operation by displaying accurate energy consumption information on products, which is useful in the purchase decision. Energy labelling of household appliances now operates in most Organization for Economic Co-operation and Development (OECD) countries and an increasing number of other countries. A wide variety of products are labelled, with the list varying from country to country. The most commonly labelled appliances are refrigerators, freezers and air conditioners, although the range is as diverse as rice cookers, boilers, lighting products and washing machines. Labelling is not restricted to electric products, with some countries including gas and oil equipment in their programs.

2.1 There are two main types of labels: endorsement and comparison

2.1.1 Endorsement labels

It indicates that products which belong to the "most energy efficient" class of products or meet a predetermined standard or eligibility criteria. Products generally display a logo or a mark which identifies they have met the standard or product class and the labels generally contains little or no comparative energy efficiency information. This type of label merely informs the consumer that the product meets the required standard. Endorsement labelling programs are mostly of a voluntary nature. An endorsement label may be specifically for energy efficiency or it may be an "Eco" label. Eco label programs endorse products that have low impact across a wide range of environmental factors, with energy consumption levels often having a high priority (but not always).

2.1.2 Comparative labels

It allows consumers to form a judgement about the energy efficiency (or energy consumption) and relative ranking of all products that carry a label. The comparative labelling programs in OECD countries are primarily mandatory, however some comparative programs in other countries are voluntary. Endorsement and comparative labels can coexist, and do so in many countries. The most commonly used comparison labels use a scale with absolutely defined efficiency categories. This type of label allows consumers to easily assess the efficiency of a product in relation to an absolute scale, by means of a simple numerical or ranking system. The concept is that it is much easier for a consumer to remember and compare a simple ranking scale (such as 1, 2, 3 or 1 star, 2 star, 3 star or A, B, C) than to remember and compare energy consumption values.

2.2 comparison labels basic types

1. Dial Label: This type of label has a "dial" or gauge, with greater efficiency linked to advancement along the gauge (more efficient represented by a clockwise arc). This type of label is used in Australia, Thailand, and Korea and is proposed for India. The number of stars or the "grading" numeral on the scale depends on the highest preset threshold for energy performance that the model is able to meet.

2. Bar Label: This type of label uses a bar chart with a grading from best to worst. All grade bars are visible on every label with a marker next to the appropriate bar indicating the grade of the model. This label is used primarily in Europe and South America. Mexico uses this concept for some products, but the format is different (although their label design is under consideration).

3. Linear Label: This label has a linear scale indicating the highest and lowest energy use of models on the market, locating the specific model within that scale. As energy is used as the comparator (rather than efficiency), it is necessary to group models into similar size categories for comparison. This model is used in North America. There are also many other energy labels that have no graphic concept to support the indication of energy efficiency – these generally rely on text to explain the efficiency or some numeric indicator of efficiency (eg EER for air conditioners, or some efficiency ranking).

3. Importance of energy label:

3.1. Importance for Retailers

- Offers basic information to clients
- Improves the quality of sales and promotion of the shop
- Positive reaction of the public due to protection of the environment
- Fulfills a legal duty

3.2. Importance for a Client

- Overall orientation of the client about energy efficiency parameters of the appliance at a time of growing energy prices
- The possibility of quick selection by comparing energy (and water) consumption between appliances in the same category
- Guarantee of identical methodology of testing and control of the content of labels

3.3. Importance for Manufacturers

- Promotion of energy efficient appliances on competitive markets
- An argument for the acceleration of replacing old inefficient appliances with modern ones
- Documentation of technological progress towards energy efficiency and the lowering of operation costs

4. Energy label schematics:

- Whilst the building regulations give guidance on how to conserve energy and power by imposing certain constraints on new buildings, there are further measures that can be taken to improve the efficiency of dwellings such as orientation, glazing strategies, internal heat transfer, ventilation controls, thermal capacity, and thermal mass (utilized with passive solar gain and in conjunction with shading and wind shelter etc.). A design guide to energy efficient housing endorses these measures, also stressing the importance of infiltration, passive solar gains, heating and hot water systems, central heating, lighting and appliances.

- Maximizing the energy efficiency of dwellings can be encouraged through the medium of energy labeling. The main commercially available labels are the National Energy Foundation`s, National Home Energy Rating (NHER) (energy efficiency rating from 0-10) and the MVM (Star point) (1-5 stars rating) scheme. The energy rating is a measure of the energy efficiency of a dwelling in terms of the annual running costs of fuel for a building, over the floor area (m2). The Government has now developed its own energy labelling scheme, the Standard Assessment Procedure (SAP), to enable labels from different systems to be compared on a common 0-100 scale.

Labels may take into account:

- Design and form of construction.
- Location.
- Site orientation and layout.
- Site exposure.
- Building form and internal layout.
- Construction details to reduce infiltration.
- The control of ventilation.
- Insulation details.
- Passive solar design.
- The selection of heating, hot water systems, lights, and appliances.
- Efficiency of heating system and controls.
- Fuel type.
- Conservatories.
- House volume.
- Number of openings and their orientation.
- Energy labels have an important role to play as an aid to designers of housing. They allow various aspects of the house design to be considered individually and their ensuing benefits assessed in cost effective priority. Energy labels can also be used as a method of assessing the efficiency of the existing housing stock. According to a recent report by Star point around 40% of U.K. homes have a two star rating and 36% have a one star rating. The philosophy at Star point has been

to create a simple system. If a house has a one Star, this indicates poor energy efficiency. Labeling systems should encourage the home owner to take measures to improve their homes.

- Energy labels are based upon Building Research Establishment Domestic Energy Model (BREDEM), and are innovative techniques available for assessing the energy efficiency of both new house construction and the existing housing stock. Energy labels can be used as a design tool to maximize the energy efficiency of housing, by allowing individual design measures to be evaluated and therefore optimized. The NHER scheme is more suitable for such applications, although more complicated and costly. The MVM scheme is quicker than the NHER system and is more suited to assessing the efficiency of the existing housing stock, suggesting the measures available to increase the rating of the property in cost-effective priority, resulting in reduced fuel consumption and ultimately lower carbon dioxide emissions. Energy labeling should encourage improved design of new housing and improvement to the existing housing stock, once people know the efficiency rating of their home.

4.1. Information Content of the Energy Label

- Name of the manufacturer and concrete product type
- Selection of energy class
- Electricity consumption for one (washing, drying) cycle or for 24 hours
- Other efficiency classes - washing, spinning, cleaning, drying, etc.
- Water consumption (washing machines, dishwashers), noise, etc.
- More, according to appliance type

4.2. Energy labeling examples:

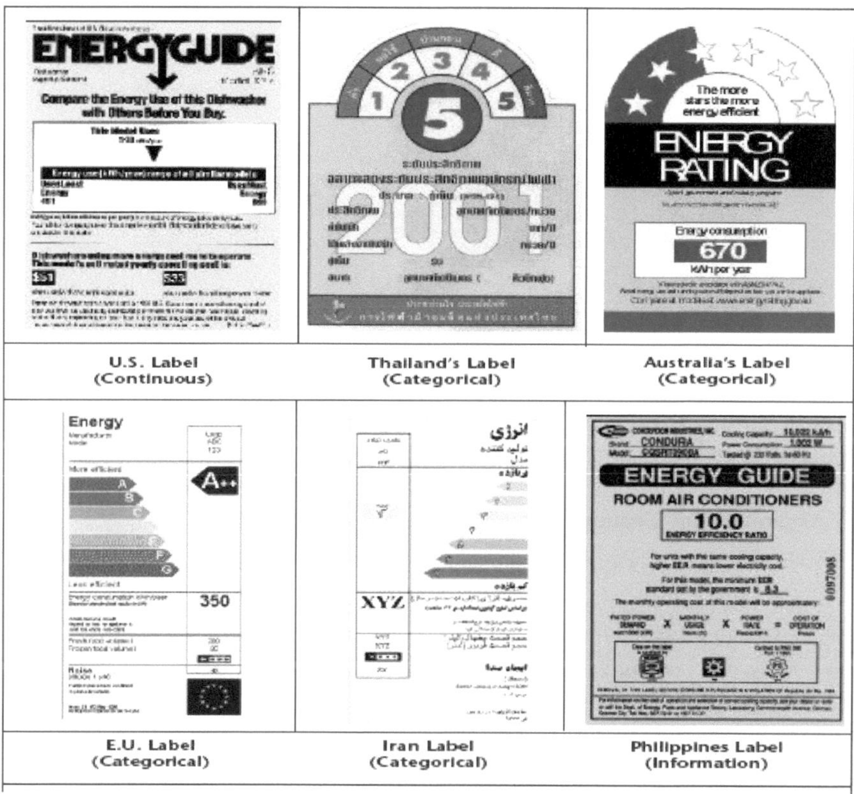

Table 1 below indicates the energy efficiency; the index is calculated for each appliance according to its consumption and its compartments' volume taking into account the appliance type. The index is thus not calculated in KW/H.

Table 1: Energy Efficiency

Energy Class	Energy Efficiency Index
A++	30> I
A+	42>I>30
A	42>I>55
B	55>I>75
C	75>I>90
D	*90>I>100*
E	*100>I>110*
F	*110>I>125*
G	*125>I*

5. Summary:

5.1. Arguments for Customers

- Growth of energy consumption and prices, more appliances in households
- Costs of appliance operation
- Average lifespan of appliances
- More efficient appliances on the market
- If all would save a little

5.2. Energy Labeling = Legal Duty

- First of all, labeling is in the consumer's best interest
- Labels and supporting documents have to be supplied by the manufacturer
- The retailer has to show the label and is responsible to do so for the customer
- Government controls label visibility and accuracy of information

5.3. Saving energy at home

- by purchasing energy efficient appliances
- by maintaining their proper upkeep and operation

CHAPTER FOUR: CALCULATION AND DESIGN PART

1. Introduction:

Most electrical devices at homes draw power even when they are not running. This is especially true with televisions, audio receivers and DVD players in a home theater system. Using an ordinary power strip for all these devices will allow this vampire power loss with the user not even knowing about it. I plan on building a power strip which will not only suppress power surges but will actually allow the user to monitor the amount of power being used, even if the devices are turned off. Also, in Auto Power Save Mode, the power strip will automatically turn off idle devices.

1.1 Objectives

The objective for this project is to develop an economical power strip that will not only protect against transient voltage surges but also end up paying for itself by reducing a user's electrical bill.

Benefits:

- Easy to use "Plug-and-Play" design will be ready to use right out of the box
- It will cut down the energy wasted by idling devices
- It will detect the power consumption

Features:

- LCD Display to show the user the power used per plug
- Safety fuses for each device to prevent the whole power strip from failing
- Simple user interface with two buttons to scroll to next plug
- Can be easily implemented to have 6 or more plugs
- Standard and Automatic Power Save Mode

1.2 Block Diagram

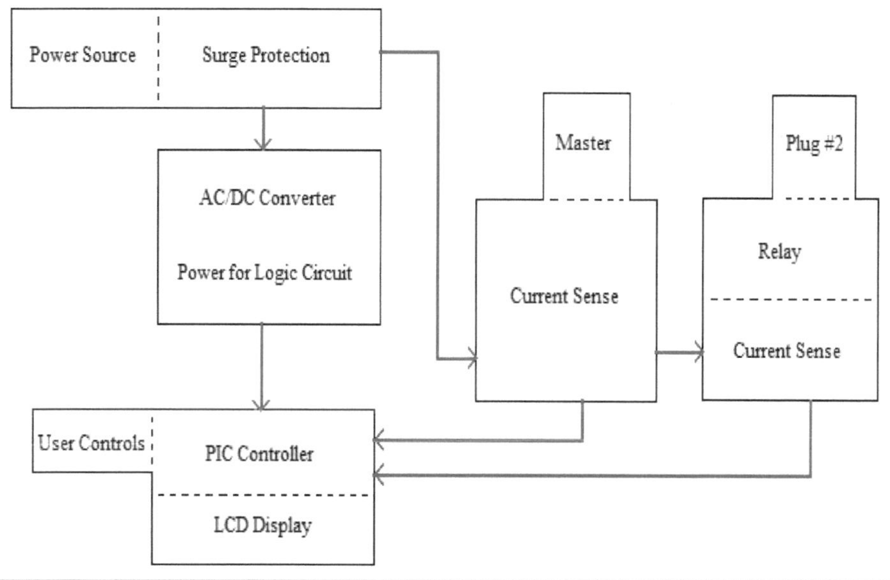

The surge protection circuit is similar to conventional power strips. There are fuses and energy dissipating elements to absorb large surges. This then feeds protected power to the AC/DC converter circuit where a steady +5VDC is created for all the CMOS logic in the project. Then, this new V_{DD} and Gnd is used by the Microcontroller and the Current Sensors. The Microcontroller is the brains of the project; this is where the user interacts with the circuit, analog calculations are made, and decisions for operation are executed. The Current Sense section is where the used current is measured and monitored.

2. Design Procedure and Details:

2.1 Schematics

Each portion from the above block diagram will be examined and explained in detail. This includes a schematic for each section along with a list of the most critical and essential parts that make that section function properly. Also there will be a description of each of these parts.

2.2 Main Protection

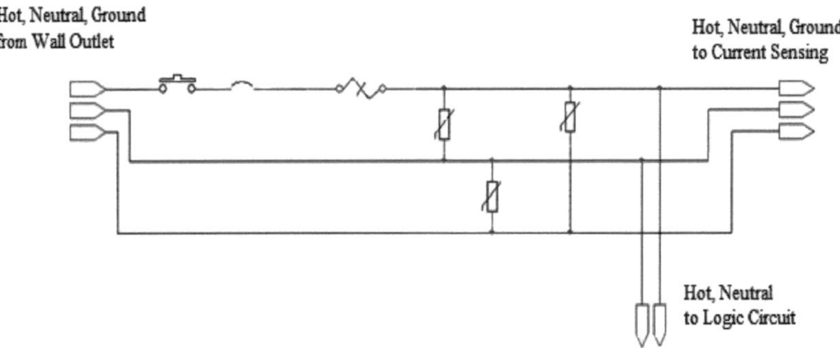

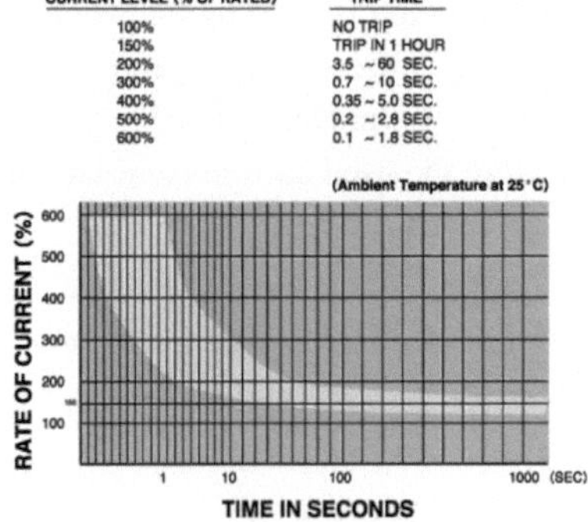

CURRENT LEVEL (% OF RATED)	TRIP TIME
100%	NO TRIP
150%	TRIP IN 1 HOUR
200%	3.5 ~60 SEC.
300%	0.7 ~10 SEC.
400%	0.35 ~ 5.0 SEC.
500%	0.2 ~2.8 SEC.
600%	0.1 ~1.8 SEC.

This portion of the project is where the main power comes in from the wall outlet. There will be about 220VAC @ 50HZ coming in and the goal is to keep it less than 15A of current so that no attached devices get ruined. In order to protect this from happening, there will first be an Illuminated Rocker Switch that has a circuit breaker built into it. The circuit breaker will trip when the rated 15A of current is sustained for an extended period of time or when the current exceeds the rated 15A. Below is a chart to show how long it will take for the circuit breaker to trip based on the current level. The datasheet is available in the Documentation section.

The next protection against high current is a 15A Thermal Fuse. This device will help prevent electrical fires by opening the circuit if the temperature exceeds the specified level of 170°C. As demonstrated below, once the thermal element melts away, the spring keeps the connection broken. Unfortunately, this device is not resettable. This device will only trip under abnormal heating due to high current flow or other malfunction that will cause an increase in temperature. This datasheet is available in the Documentation section

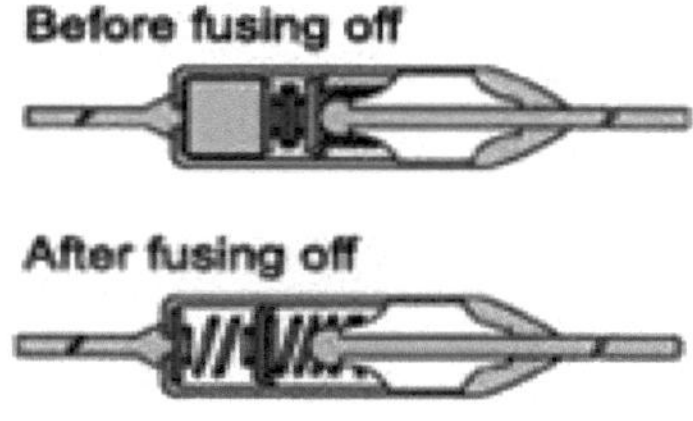

The surge protection element is the Metal Oxide Varistor. This device is designed to absorb large voltage surges and dissipate the energy absorbed. This part acts like an open circuit as long as the voltage across it is less than the threshold voltage. Once the threshold voltage is exceeded it begins to conduct current and absorb as much energy as possible. The manufacturer suggests that this device will act as an open circuit as long as the voltage remains less than 200VAC. Once that is exceeded, the varistor will begin conducting current and dissipating energy. It will survive a surge current of 250A. It is suggested that each varistor will dissipate 300J of energy; with three of these elements connected as shown in the schematic above, this power strip will have surge suppression of over 900J which is far superior to the conventional 150J-300J that most power strips can handle.

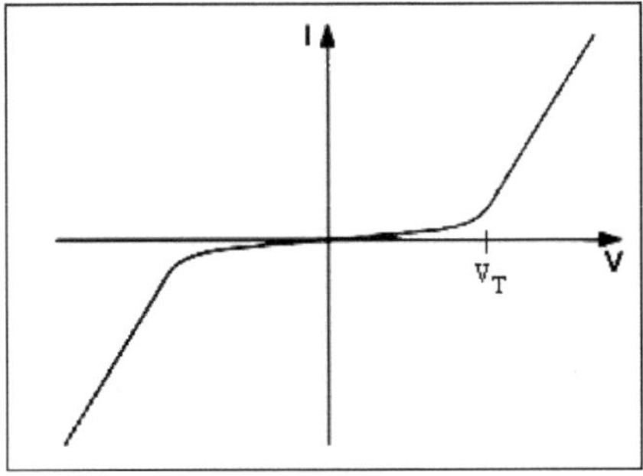

2.3 Step Down / AC to DC

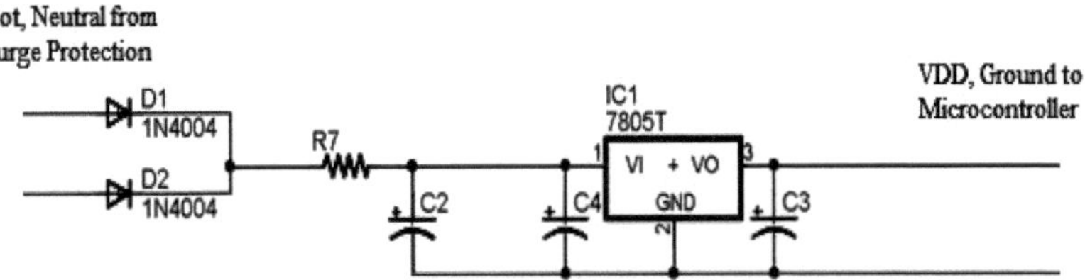

First, to step-down the voltage to a safe V_{DD}, a 12.6VAC Center-Tapped 2.0A Transformer will be used. The center will become the local ground for the remainder of the circuit. This is not to be confused with the earth ground that the wall outlet provides. This ground will just be the local 0V that all the TTL and CMOS devices will use. Then, the AC wave is rectified with two 1N4001 diodes in a full-wave rectifying setup. With each diode having a voltage drop of about 0.8V, the voltage out of the rectifier peaks at around 10.7V (this is verified in the Testing section).

Rectifier Output = 10.7V, Desired Input Voltage = 8.0V

10.7-8.0 = 2.7V ➔ $V_{r(pp)} = 2.7V$

$V_{r(pp)} = \frac{V_{p(in)}T}{RC}$ ➔ $2.7 = \frac{10.7}{120(RC)}$

C = 100μF, R = 330Ω

To make sure the ripple doesn't fall below +5V ,we chose the maximum V_{ripple} to be no greater that 2.7V to go into the voltage regulator. Inserting a 100μF capacitor steadies the oscillations and insures that the ripple never g0oes below 8.4V (verified in the Testing section). In order for the MC7805 Voltage Regulator to output a steady and smooth +5V, the input voltage is recommended by the manufacturer to be between 7V and 20V. By having a maximum ripple from 8.4V to 10.7V this guarantees that the circuit is well into the operating range for the Voltage Regulator (verified in the Testing section). The manufacturer also suggests connecting the Voltage Regulator with two capacitors to help reduce any noise going into the input and coming out of the output. Now, the output becomes V_{DD} and Gnd for the rest of digital logic.

2.4 PIC Microcontroller

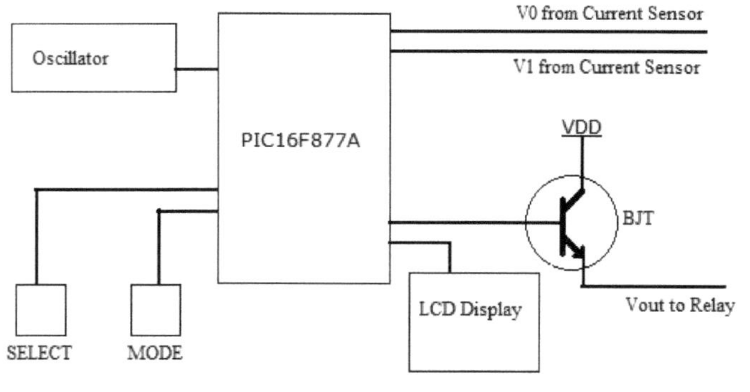

The first step for the PIC16F877A will be to convert analog signals V_0 and V_1 from the current sensors into digital signals. The analog signals are fed as inputs into the PIC via pins AN0 and AN1 respectively. Then, it must take those digital signals and quantify them into 8-bit values to determine the current flowing to the plugs and do some arithmetic to calculate the power consumption per plug. The result was an integer between 0 and 255 from the A/D conversion. We want the result to be in mA so this calculation will give that result.

$$mA = \frac{\left|\left(\left(\frac{ADvalue}{255}\right) * 5000\right) - 2500\right|}{0.180}$$

A quick explanation of this calculation is necessary. First, take the ADvalue and divide it by 255 and multiply by 5000 to give the mV that is being read. Now, when there is 0mA of current, the sensor outputs 2.5V (2500mV). Subtract that to give you the zero point. Then take the absolute value to account for positive or negative current. Finally, since the current sensor has a linear output relationship of 180mV is 1A, you divide by .180 to give the number of mA.

Next, the PIC will have two de-bounced buttons as inputs into pins C_0 and C_1. These will be the SELECT and MODE buttons that the user will press to select between plugs or modes of operation. With the values of R and C 10kΩ and 10nF, this gives a time constant of τ=RC which is 0.1ms; this will be enough time to settle down any bouncing that may occur while being pressed. The recommended oscillator frequency for this PIC is 20MHz. The chosen oscillator was a 20MHz Crystal Oscillator (operation was verified in the Testing section) and this is connected to the external clock/oscillator pins 13 and 14. The 16x2 LCD Display will be driven from Port D because this port has the RS232 serial output capability. The output going to the BJT and the relay is the signal which determines if the second plug will be disabled or not when in Auto Power Save Mode. The minimum coil current in the relay for the current sense section is 50mA to activate the switch. The PIC can only source about 25mA so this simple NPN Bipolar Junction Transistor circuit was created. The PIC now only needs to provide 2mA so that the BJT can turn on and source the current for the relay. This particular transistor has a gain β=10000. $I_C = \beta * I_B$ for a BJT transistor it is easy to see that a minimal base current from the PIC is necessary to allow the 50mA to flow through the BJT.

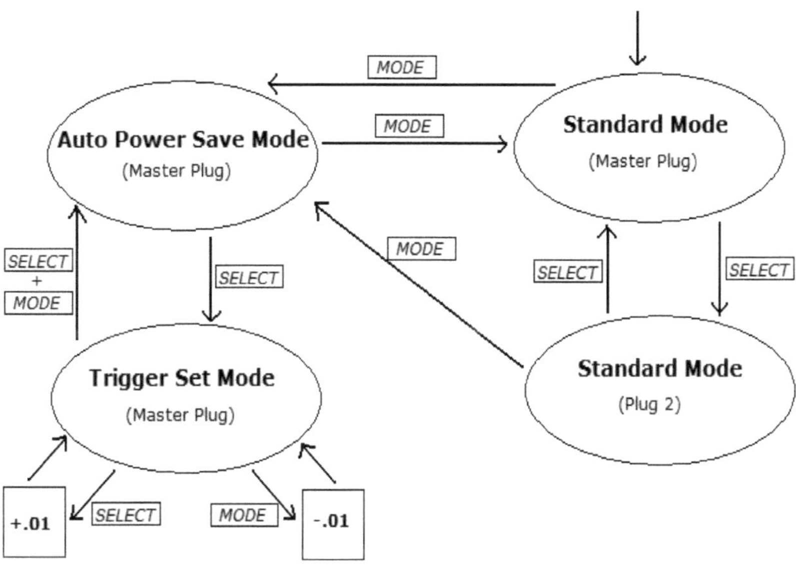

Above is a flow chart for the programming of the PIC. It starts in Standard Mode where all the plugs are always on. The user can press SELECT to cycle through the plugs and the LCD will display instantaneous current. Pressing MODE will take the user to Auto Power Save Mode where the first thing that happens is a threshold value is set from the Master plug. Now while in this mode, if the current of the Master plug falls below 70% of the set threshold, plug 2 will automatically be disabled until the current return above the threshold. 70% is an acceptable threshold for most devices but if the user has a device where this threshold is unacceptable, the user can press the SELECT button to bring them into the Trigger Set Mode. Here, the LCD displays the current trigger point and the user can press SELECT or MODE to increase or decrease the threshold manually. Once the user is satisfied, pressing SELECT and MODE together will save the new threshold and return to Auto Power Save. Pressing the MODE button here will take the user back to Standard Mode where all plugs will be enabled.

2.5 Current Sense

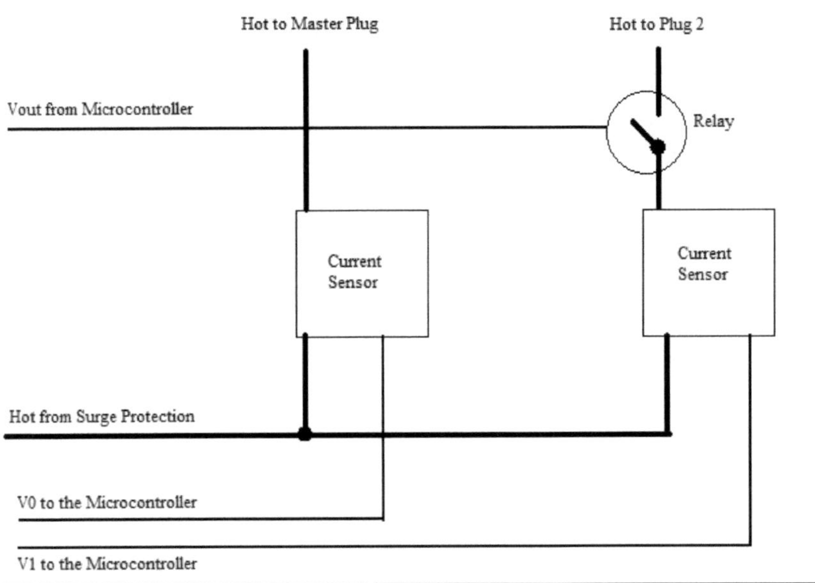

This portion of the smart power strip is arguably the most important section. I have found an IC which uses a Hall-Effect sensor to produce a linearly related analog voltage based on the current measured. The ACS714 is rated up to 5A of AC current but can safely operate beyond that as long as heat is controlled. The graph below was acquired from the manufacturer's datasheet and shows the linear relationship of current sensed and output voltage.

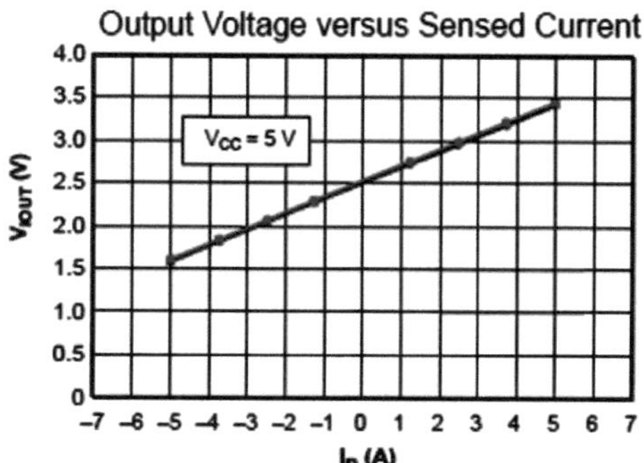

There are data points which verify this linear relationship in the testing and verification section. The analog output of the IC – V_0, V_1 – will be fed into the PIC Microprocessor for each plug. The IC will use a V_{DD} and Ground that come from the Step-Down and AC/DC circuit. This logic 0V should not be confused with the earth ground obtained from the wall outlet that will be used for the plugs. Each plug will be the US standard 220VAC grounded outlet. Plug 2 will have a relay in series enabling/disabling the plug based on the calculations from the microcontroller. The minimum voltage across the relay coil (to active it) is 3.5V. With and internal resistance of 70Ω, that takes a minimum current of 50mA to activate the switch.

3. Verification and testing:

3.1 Main Protection

Because high current and high voltage surges are dangerous and difficult to do, I have decided to trust the manufacturers' values for the amount of protection these elements offer. Also, the configuration in which I have connected these elements is nothing new; this schematic is widely used and accepted by many power strip manufacturers. Instead of testing the limitations and failure points of these elements, I will instead, test the normal operations to make sure that there is very little leakage current and to monitor temperatures of all the elements. As a result at the normal operating voltage of 220VAC at 50HZ, there is no sign of the varistors.

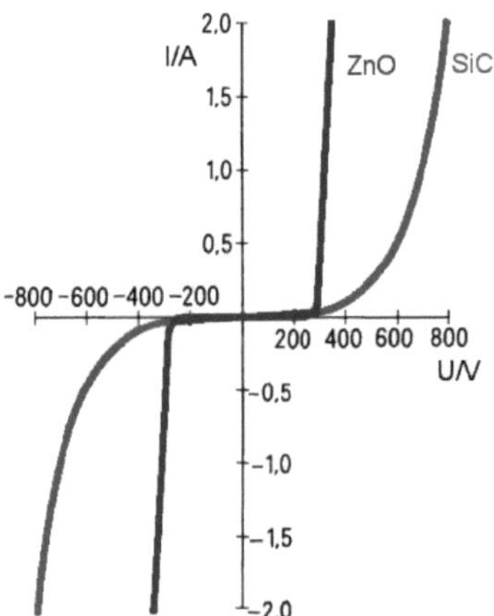

Two waveforms were examined side by side, the voltage with no varistors, no switches, and no thermal fuse. Then, the same waveform was examined after adding three varistors, the main Illuminated Rocker Switch, and the 15A Thermal Fuse and the two were identical. This shows that these components have no interference with circuit except to protect it in case of abnormal operation or failure.

3.2 Step Down / AC to DC

I have begun assembling and testing parts of this circuit already. First was to measure the output of the transformer. Using the oscilloscope I connected the ground to the Center-Tapped lead and measured the oscillating output of the transformer.

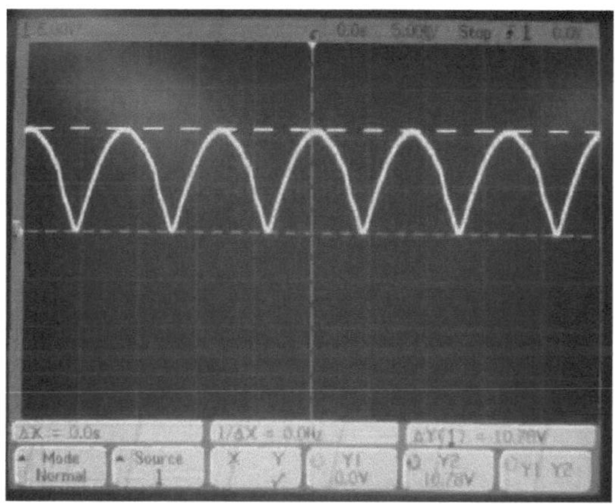

The resulting voltage peaks were +11.5V to -11.5V. Then, the voltage was rectified using two diodes in a full-wave rectifier configuration and the oscilloscope was used again to see the resulting voltage peaks. Because there is about a 0.8V drop across the diodes, the resulting peaks were almost 10.7V.

Once that verified, it was time to find the maximum V_{ripple} allowable to not interfere with the voltage regulator. Standard operating input voltage is 7V to 20V according to the Manufacturer; this meant that we didn't want the bottom of the ripple to go below 8V to make sure it didn't approach the minimum 7V. Using a 100µF capacitor insured that the voltage didn't go below 8.4V. It stayed between 10.78V and 8.44V.

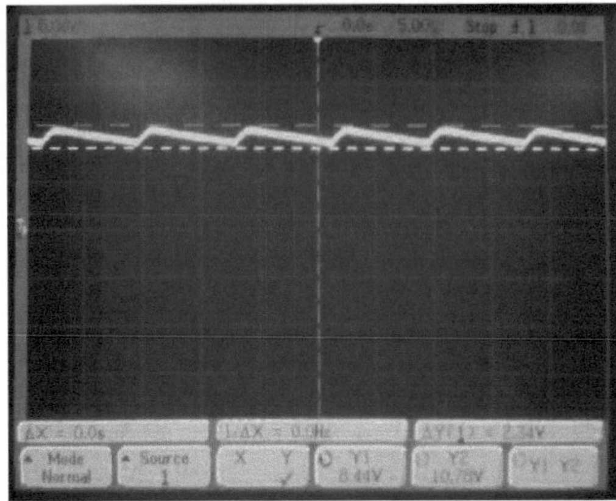

That ripple was then fed into the Voltage Regulator which should have produced a steady +5V level for the TTL and CMOS circuitry to use. Below is verification that the regulator did in fact produce a steady +5V level. The level stays between 4.7V and 5.5V which is a solid logic high for the rest of the circuit to use.

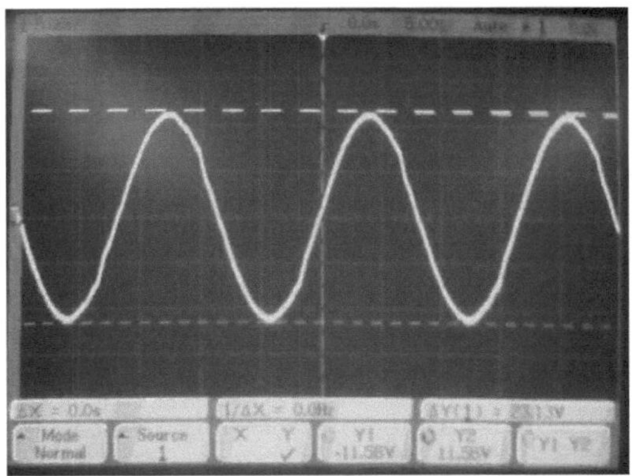

3.3 Current Sense

The most important component for this section to test was the current sensor to verify that the output relationship is indeed linear. The circuit which was constructed was just a power supply and a 5mΩ resistor so that a high current can be passed through.

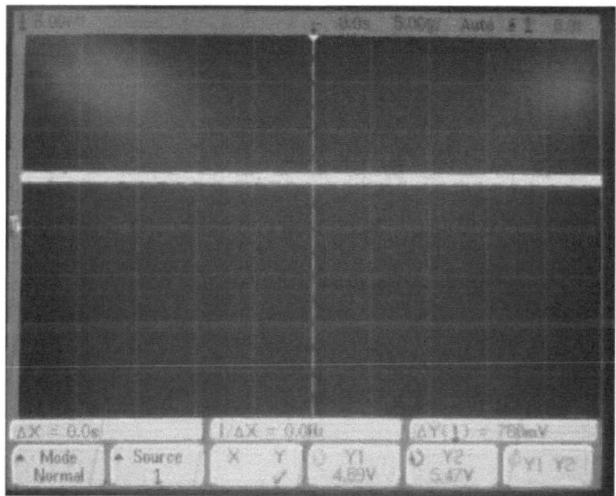

The resistor was capable of handling up to 5 Watts. Then, we just added the current sensor in series and a voltmeter connect to the V_{out} of the sensor and recorded data points.

4. Conclusions:

4.1 Accomplishments

I feel that I have designed and created a complete project which was functioning at the end with only minor problems. I think I have succeeded in building a smart power strip like according to what I had proposed. Its ability to monitor power consumption is fairly accurate; especially the greater the current.

4.2 Uncertainties

I had a lot of difficult calibrating the PIC due to AC noise on PCB. The noise mainly affected the current sensors and the LCD. The current sensors had a lot of fluctuations around the values they should have been displaying. In the future I would create a separate PCB for the DC components and a separate PCB for the AC components, that way it would reduce the noise in the system.

5. References

1. Electricity 101/slides for Edison electric institute.

2. Energy conservation /pollution prevention (p2) education toolbox/tools for helping teachers integrate p2 concepts in the classroom / United States environmental protection agency/ epa-905-f-97-011 august 1997.

3. Energy conservation methods/ general guidelines.

4. Monitoring changes in economy-wide energy efficiency: from energy–gdp ratio to composite efficiency index/b.w. Ang.

5. A literature review on energy efficiency standards and labels for household electrical appliances/masjuki h.h., t. M. I. Mahlia*, i. A. Choudhury, and. Saidur.

6. Energy labeling and standards programs throughout the world report prepared by Lloyd Harrington energy efficient strategies, Australia and Melissa damnics energy consult, Australia.

7. Who needs energy labelling? An assessment of energy efficiency measures adopted in UK housing in response to perceived client need /Dr. Hilary Davies the Hong Kong polytechnic university.

8. Energy labeling of household appliances and its importance to retailers.

9. Smart strip power strip model overviews/bitsltd.net.

10. Approach technology Inc. Available at: http://www.apprtech.com.tw/ss-002.htm.

11. Automotive grade, fully integrated, hall effect-based linear current sensor ic with 2.1 kvrms voltage isolation and a low-resistance current conductor. Available at: http://www.allegromicro.com/en/Products/Current-Sensor-ICs/Zero-To-Fifty-Amp-Integrated-Conductor-Sensor-ICs/ACS714.aspx (last accessed July 6, 2016).

12. Cantherm. Available at: http://www.cantherm.com/products/thermal-fuses.html

13.All data sheet (Mc7805). Available at:
http://www.alldatasheet.com/view.jsp?Searchword=Mc7805

14.Engineering at Illinois, Engineering IT, Shared Services. Available at:
https://courses.engr.illinois.edu/ece395/docs/clock%20f1100e.pdf

15.On Semiconductor, Energy Efficient Innovations, Axial Lead Standard Recovery Rectifiers. Available at:
http://www.onsemi.com/pub_link/Collateral/1N4001-D.PDF

16.World Products. Available at:
http://www.worldproducts.com/pdfs/ProtectionProducts/WPMOV_Rev10.7.pdf

17.http://en.wikipedia.org/

18.http://www.chipswinner.com/DS/MC7805.pdf

Printed by Books on Demand GmbH, Norderstedt / Germany